Introduction to Weatherization

Trainee Guide
First Edition

Prentice Hall

Boston Columbus Indianapolis New York San Francisco Upper Saddle River
Amsterdam Cape Town Dubai London Madrid Milan Munich Paris Montreal Toronto
Delhi Mexico City Sao Paulo Sydney Hong Kong Seoul Singapore Taipei Tokyo

National Center for Construction Education and Research

President: Don Whyte
Director of Product Development: Daniele Stacey
Introduction to Weatherization Project Manager: Jennifer Wilkerson
Production Manager: Tim Davis
Quality Assurance Coordinator: Debie Ness
Editor: Rob Richardson
Desktop Publishing Coordinator: James McKay
Production Assistant: Laura Wright
Cover Photo: Tim Davis with the assistance of Jennifer Wilkerson/NCCER

NCCER would like to acknowledge the contract service provider for this curriculum:
Topaz Publications, Syracuse, New York.

This information is general in nature and intended for training purposes only. Actual performance of activities described in this manual requires compliance with all applicable operating, service, maintenance, and safety procedures under the direction of qualified personnel. References in this manual to patented or proprietary devices do not constitute a recommendation of their use.

Copyright © 2010 by the National Center for Construction Education and Research (NCCER) and published by Pearson Education, Inc., publishing as Prentice Hall. All rights reserved. Manufactured in the United States of America. This publication is protected by Copyright, and permission should be obtained from NCCER prior to any prohibited reproduction, storage in a retrieval system, or transmission in any form or by any means, electronic, mechanical, photocopying, recording, or likewise. To obtain permission(s) to use material from this work, please submit a written request to NCCER Product Development, 3600 NW 43rd St., Building G, Gainesville, FL 32606.

Many of the designations by manufacturers and sellers to distinguish their products are claimed as trademarks. Where those designations appear in this book, and the publisher was aware of a trademark claim, the designations have been printed in initial caps or all caps.

10 9 8 7 6 5 4 3 2

ISBN 13: 978-0-13-216699-7

www.pearsonhighered.com

Preface

To the Trainee

The weatherization industry is a rapidly growing field that encourages homeowners to invest in improving the energy efficiency of their homes. The retrofits involved in weatherization can reduce energy waste while improving the comfort and safety within a house. In addition, this type of energy conservation helps our country reduce its dependence on foreign oil and decrease greenhouse gas emissions.

The new emphasis in this field, along with government initiatives, has created an overwhelming demand for qualified weatherization craft workers. As of the date of this publication, $5 billion has already been allotted for weatherization measures performed on qualified American households, and additional programs are projected to add another $6 billion. With an estimated 128 million qualified homes and an expected $19 billion-per-year market, the time for training is now.

This module covers a broad range of information that is vital to everyone working in the weatherization field, including an overview of how to identify and seal air leaks, locate inadequate insulation, and reduce baseloads. *Introduction to Weatherization* is intended to introduce trainees to the concepts and skills they will need to be successful in *Weatherization Technician, Level One*.

Weatherization Fundamentals

Combined with NCCER's *Core Curriculum*, the *Introduction to Weatherization* makes up *Fundamentals of Weatherization*. If you're training through an NCCER-Accredited Training Program Sponsor, and you've successfully passed the module exams and performance tests in this course, you may be eligible for credentialing through NCCER's National Registry. Check with your instructor or local program sponsor to find out. To learn more, go to www.nccer.org or contact us at 1.888.622.3720.

We invite you to visit the NCCER website at www.nccer.org for the latest releases, training information, newsletter, and much more. You can also reference the Contren® product catalog online at www.nccer.org.

Your feedback is welcome. You may email your comments to curriculum@nccer.org or send general comments and inquiries to info@nccer.org.

Contren® Learning Series

The National Center for Construction Education and Research (NCCER) is a not-for-profit 501(c)(3) education foundation established in 1995 by the world's largest and most progressive construction companies and national construction associations. It was founded to address the severe workforce shortage facing the industry and to develop a standardized training process and curricula. Today, NCCER is supported by hundreds of leading construction and maintenance companies, manufacturers, and national associations. The Contren® Learning Series was developed by NCCER in partnership with Pearson Education, Inc., the world's largest educational publisher.

Some features of NCCER's Contren® Learning Series are as follows:

- An industry-proven record of success
- Curricula developed by the industry for the industry
- National standardization providing portability of learned job skills and educational credits
- Compliance with Office of Apprenticeship requirements for related classroom training (CFR 29:29)
- Well-illustrated, up-to-date, and practical information

NCCER also maintains a National Registry that provides transcripts, certificates, and wallet cards to individuals who have successfully completed modules of NCCER's Contren® Learning Series. *Training programs must be delivered by an NCCER Accredited Training Sponsor in order to receive these credentials.*

Contren® Curricula

NCCER's training programs comprise more than 80 construction, maintenance, pipeline, and utility areas and include skills assessments, safety training, and management education.

Boilermaking
Cabinetmaking
Carpentry
Concrete Finishing
Construction Craft Laborer
Construction Technology
Core Curriculum:
    Introductory Craft Skills
Drywall
Electrical
Electronic Systems Technician
Heating, Ventilating, and
    Air Conditioning
Heavy Equipment Operations
Highway/Heavy Construction
Hydroblasting
Industrial Coating and Lining
    Application Specialist
Industrial Maintenance
    Electrical and
    Instrumentation Technician
Industrial Maintenance
    Mechanic
Instrumentation
Insulating
Ironworking
Masonry
Millwright
Mobile Crane Operations
Painting
Painting, Industrial
Pipefitting
Pipelayer
Plumbing
Reinforcing Ironwork
Rigging
Scaffolding
Sheet Metal
Site Layout
Sprinkler Fitting
Tower Crane Operator
Welding

Green/Sustainable Construction
Your Role in the Green Environment
Introduction to Weatherization
Fundamentals of Weatherization
Weatherization Technician
Weatherization Crew Chief
Energy Auditor
Sustainable Construction Supervisor

Energy
Introduction to the Power Industry
Power Industry Fundamentals
Power Generation Maintenance Electrician
Power Generation I&C Maintenance Technician
Power Generation Maintenance Mechanic
Steam and Gas Turbine Technician
Introduction to Solar Photovoltaics
Introduction to Wind Energy

Pipeline
Control Center Operations, Liquid
Corrosion Control
Electrical and Instrumentation
Field Operations, Liquid
Field Operations, Gas
Maintenance
Mechanical

Safety
Field Safety
Safety Orientation
Safety Technology

Management
Introductory Skills for the Crew Leader
Project Management
Project Supervision

Supplemental Titles
Applied Construction Math
Careers in Construction
Tools for Success

Spanish Translations
Basic Rigging
    (Principios Básicos de Maniobras)
Carpentry Fundamentals
    (Introducción a la Carpintería, Nivel Uno)
Carpentry Forms
    (Formas para Carpintería, Nivel Trés)
Concete Finishing, Level One
    (Acabado de Concreto, Nivel Uno)
Core Curriculum:
    Introductory Craft Skills
    (Currículo Básico: Habilidades Introductorias del Oficio)
Drywall, Level One
    (Paneles de Yeso, Nivel Uno)
Electrical, Level One
    (Electricidad, Nivel Uno)
Field Safety
    (Seguridad de Campo)
Insulating, Level One
    (Aislamiento, Nivel Uno)
Masonry, Level One
    (Albañilería, Nivel Uno)
Pipefitting, Level One
    (Instalación de Tubería Industrial, Nivel Uno)
Reinforcing Ironwork, Level One
(Herreria de Refuerzo, Nivel Uno)
Safety Orientation
    (Orientación de Seguridad)
Scaffolding
    (Andamios)
Sprinkler Fitting, Level One
    (Instalación de Rociadores, Nivel Uno)

Acknowledgments

This curriculum was revised as a result of the farsightedness and leadership
of the following sponsors:

AGC of Oklahoma – Building Chapter
Entek Corporation
McNeal and White Contractors
National Association of Minority Contractors

North American Insulation Manufacturers
Pennsylvania College of Technology
Tallahassee Community College

This curriculum would not exist were it not for the dedication and unselfish energy of those volunteers
who served on the Authoring Team. A sincere thanks is extended to the following:

Steve Buglione
Rick Frazier
Larry Leonard
John Manz

JR McNeal
Howard Smith
Matthew Todd
Darrell Winters

NCCER Partners

American Fire Sprinkler Association
Associated Builders and Contractors, Inc.
Associated General Contractors of America
Association for Career and Technical Education
Association for Skilled and Technical Sciences
Carolinas AGC, Inc.
Carolinas Electrical Contractors Association
Center for the Improvement of Construction Management and Processes
Construction Industry Institute
Construction Users Roundtable
Design Build Institute of America
Manufacturing Institute
Merit Contractors Association of Canada
Metal Building Manufacturers Association
NACE International
National Association of Minority Contractors
National Association of Women in Construction
National Insulation Association
National Ready Mixed Concrete Association
National Technical Honor Society

National Utility Contractors Association
NAWIC Education Foundation
North American Technician Excellence
Painting & Decorating Contractors of America
Portland Cement Association
SkillsUSA
Steel Erectors Association of America
U.S. Army Corps of Engineers
University of Florida
Women Construction Owners & Executives, USA

Contents

**59101-10
Introduction to
Weatherization............. 1.i**

This module introduces the trainee to the weatherization initiative and its underlying motivation by examining the economic and environmental effects of the inefficient use of energy in heating and cooling buildings. The module describes the common ways in which heat is lost and how cold air infiltrates a house. It introduces remediation methods such as air sealing and insulation as a lead-in to subsequent modules. The module also explores career opportunities in the weatherization industry.

Index I.1

Training Opportunities within NCCER's Weatherization Program

NCCER's Weatherization Program offers the trainee several different options. After successfully completing the *Fundamentals of Weatherization* and *Weatherization Technician, Level One*, the trainee can enter the industry as a Weatherization Technician. If the trainee continues in the program, he/she can choose between instruction for a Weatherization Crew Chief or for an Energy Auditor. Either option will advance the skills of the trainee. The decision should be based on what career in the weatherization industry the trainee is most interested in.

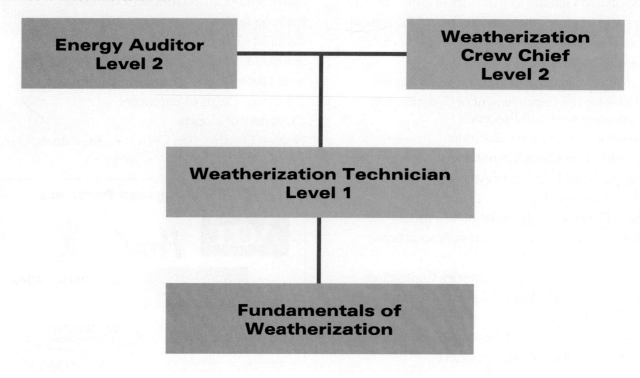

Introduction to Weatherization

59101-10

Introduction to Weatherization

Objectives

When you have completed this module, you will be able to do the following:

1. Explain the purpose, benefits, and origin of the home weatherization program.
2. Explain how weatherization goals are met by reducing heating and cooling losses and by reducing air infiltration.
3. Describe how sources of heating and cooling losses and air infiltration points are located.
4. Describe the methods and materials used to reduce heating and cooling losses and to stop air infiltration.
5. Describe how the different components that make up the building shell can affect a home's energy usage.

Trade Terms

Baseload
Blower door
British thermal unit (Btu)
Compact fluorescent lamp
Conditioned air
Cubic feet per minute (cfm)
Energy Star®
Equilibrium
Exfiltration
Heat gain
Heat loss
Infiltration

Infrared camera
Light-emitting diode (LED)
Low-E glass
Radon
Return air duct
Rim joist
R-value
Stack effect
Supply air duct
U-value
Vapor barrier

Fundamentals of Weatherization course map:

- 00109-09 Introduction to Materials Handling
- 00108-09 Basic Employability Skills
- 00107-09 Basic Communication Skills
- 00106-09 Basic Rigging
- 00105-09 Introduction to Construction Drawings
- 00104-09 Introduction to Power Tools
- 00103-09 Introduction to Hand Tools
- 00102-09 Introduction to Construction Math
- 00101-09 Basic Safety
- 59101-10 Introduction to Weatherization

This course map shows all of the modules in *Fundamentals of Weatherization*. The suggested training order begins at the bottom and proceeds up. Skill levels increase as you advance on the course map. The local Training Program Sponsor may adjust the training order.

Contents

Topics to be presented in this module include:

1.0.0 Introduction 1.1
2.0.0 Weatherization Concepts 1.1
   2.1.0 Home Health and Safety 1.1
   2.2.0 Equipment Condition 1.1
   2.3.0 Tightness of the Building Shell 1.2
   2.4.0 Home Lighting 1.2
   2.5.0 Heat Loss and Heat Gain 1.2
   2.6.0 Air Infiltration 1.4
3.0.0 Finding Air Leaks and Inadequate Insulation 1.7
   3.1.0 Visual Inspection of the Home 1.8
   3.2.0 Finding Air Leaks 1.9
      3.2.1 Finding Air Leaks With a Blower Door 1.12
      3.2.2 Finding Inadequate Insulation 1.12
4.0.0 Weatherizing a Home 1.14
   4.1.0 Adding Insulation 1.14
      4.1.1 Types of Insulation 1.14
      4.1.2 Flexible Insulation 1.15
      4.1.3 Rigid Foam Board 1.15
      4.1.4 Loose-Fill Insulation 1.16
      4.1.5 Spray-in-Place Insulation 1.16
      4.1.6 Spray Foam Insulation 1.17
   4.2.0 Sealing Air Leaks 1.17
      4.2.1 Caulks and Sealants 1.18
      4.2.2 Weatherstripping 1.18
   4.3.0 Losses Through Windows and Doors 1.18
      4.3.1 Upgrading Windows and Doors 1.20
      4.3.2 Replacement Windows 1.20
      4.3.3 Replacement Doors 1.21
   4.4.0 Energy-Efficient Roofs 1.21
   4.5.0 Sealing and Insulating Air Ducts 1.21
5.0.0 Reducing the Baseload 1.21
   5.1.0 Appliances 1.22
      5.1.1 Refrigerators 1.22
      5.1.2 Other Appliances 1.23
      5.1.3 Water Heaters 1.23
   5.2.0 Lighting 1.23
   5.3.0 The Energy Auditor as an Educator 1.24
6.0.0 Careers in Weatherization 1.25
   6.1.0 Weatherization Technician 1.25
   6.2.0 Weatherization Crew Chief 1.25
   6.3.0 Energy Auditor 1.26
   6.4.0 Standardized Training by NCCER 1.26
   6.5.0 Advancement Opportunities In Weatherization 1.27
Appendix Samples of NCCER Training Credentials 1.33

Figures and Tables

Figure 1 Temperature equilibrium 1.3
Figure 2 Heat loss 1.3
Figure 3 Heat gain 1.4
Figure 4 Infiltration and air leaks 1.5
Figure 5 Stack effect 1.6
Figure 6 Simplified ideal air duct system 1.6
Figure 7 Simplified ideal air duct system with leaks in the supply duct ... 1.6
Figure 8 Simplified ideal air duct system with leaks in the return duct 1.7
Figure 9 Air leaks through cracks or open seams 1.9
Figure 10 Air leaks around pipes and wires 1.9
Figure 11 Air leaks around a can light 1.9
Figure 12 Air leaks around an attic access door 1.9
Figure 13 Attic air leaks 1.10–1.11
Figure 14 Blower door installed in an exterior door 1.12
Figure 15 Finding air leaks with a blower door 1.13
Figure 16 Finding an air leak 1.13
Figure 17 Infrared camera 1.13
Figure 18 Conventional and infrared images 1.14
Figure 19 Different types of insulation 1.15
Figure 20 Flexible insulation 1.15
Figure 21 Rigid foam board 1.16
Figure 22 Closure board made of rigid foam board 1.16
Figure 23 Applying loose-fill insulation 1.16
Figure 24 Applying spray-in-place insulation 1.17
Figure 25 Portable spray foam kits 1.17
Figure 26 Sealing an air leak with spray foam 1.17
Figure 27 Apply caulk with a caulking gun 1.18
Figure 28 Sealing air leaks around a double-hung window 1.18
Figure 29 Sealing air leaks around sliding windows 1.19
Figure 30 Sealing air leaks at the bottom of a door 1.20
Figure 31 Effect of glass on heating and cooling loads 1.20
Figure 32 Double- and triple-pane windows 1.21
Figure 33 Duct joints sealed with mastic 1.21
Figure 34 Seasonal and baseload energy use 1.22
Figure 35 Water heater wrapped to prevent heat loss 1.24
Figure 36 Compact fluorescent lamps 1.24

Table 1 U-Values of Window Glass 1.21

1.0.0 INTRODUCTION

Much of the energy used to heat and cool homes in the United States is wasted. This is especially true with older homes that were built at a time when energy was cheap and little thought was given to conserving it. Times have changed. Energy is now expensive and the fuels such as oil and gas used to produce it are becoming scarce. Add the fact that burning these fuels can add air pollution, and it is clear that steps must be taken to conserve energy.

Energy auditors and weatherization technicians are on the front line of this battle. When a home is weatherized, the homeowner gains increased comfort and can save hundreds of dollars a year in energy costs. As a technician, you gain directly by providing a valuable service and the knowledge that your skills provide you with job security. The nation gains by reducing its dependence on foreign fuel and by having cleaner air.

The key to proper home weatherization is the energy audit. The energy auditor inspects the home to determine areas where energy is being wasted by using test equipment such as a *blower door* or an *infrared camera*. The auditor then prepares a work order describing what the weatherization crew must do to seal the home and reduce energy use. The energy auditor also advises the homeowner about other ways that energy can be saved in the home.

The U.S. government recognizes the benefits of home weatherization. The U.S. Department of Energy (DOE) has supported this effort for many years. Its Weatherization Assistance Program (WAP) helps low-income families improve the efficiency of their homes. Weatherization programs are often funded by the federal government but are implemented by state and local governments and community action agencies. For that reason, programs will vary across the country, with each locality having its own special requirements.

The *American Recovery and Reinvestment Act (ARRA) of 2009* provides additional funds for states for WAP and provides tax credits for homeowners who upgrade their homes to make them more energy-efficient. The following home improvements qualify for tax credits to applicable households:

- Biomass stoves
- Heating and air conditioning equipment
- Insulation
- Metal and asphalt roofs
- Water heaters (non-solar)
- Windows and doors
- Solar energy systems
- Residential wind turbines

2.0.0 WEATHERIZATION CONCEPTS

The weatherization process begins and ends with an energy audit. To better understand how and where to weatherize a home to prevent energy losses, it is important to understand that an energy audit looks at seasonal energy use and *baseload*. Seasonal energy use is the energy used to heat and cool the home. Baseload is defined as the energy used in the home for other uses such as lighting and operating appliances. Items that are checked and inspected during an energy audit include the following:

- Home health and safety
- Condition and age of the furnace or boiler, water heater, air conditioner, and major appliances
- Tightness of the building shell
- Home lighting

In addition to these items, the energy auditor also looks for other areas in the home where energy savings can be had. He or she can then advise the homeowner of simple and low-cost steps that can save energy in those other areas of the home.

2.1.0 Home Health and Safety

A home with problems that could affect the health and safety of the occupants and/or the weatherization crew must have those problems identified by the auditor and corrected before the home can be weatherized. Typical health and safety issues of this nature include the following:

- Roof or basement leaks that could damage insulation and cause mold or rot
- Structural defects
- The presence of asbestos or vermiculite insulation
- The presence of lead paint
- A defective chimney or vent used with a furnace or water heater
- A cracked furnace heat exchanger
- The presence of carbon monoxide (CO)
- Incorrectly installed furnace, stove, or appliance
- Unsafe wiring
- Plumbing leaks
- Obvious building code violations

> **NOTE**
>
> Effective April 2010, the U.S. Environmental Protection Agency (EPA) requires that all individuals involved in interior and exterior renovation work be certified under the Lead Renovation, Repair, and Painting rule to ensure that they follow specific lead-safe work practices.

On-Site

Lead-Safe Training

Weatherization technicians will work in older (pre-1978) homes that contain lead paint. The EPA now requires that all individuals and businesses involved in interior and exterior renovation work be certified under the Lead Renovation, Repair and Painting rule to ensure that they follow specific lead-safe work practices. One-day certification courses are available throughout the country.

2.2.0 Equipment Condition

Part of the energy audit is to determine the age and condition of the major appliances, furnace or boiler, water heater, and air conditioner (if any) in the home. Older or poorly maintained equipment uses more energy than modern equipment. Age and condition issues include the following:

- A furnace with a cracked heat exchanger
- An oil burner that is not a flame-retention type
- Furnace or boiler combustion efficiency
- Condition of forced-air ducts or water pipes in a hot water heating system
- Condition of the furnace air filter
- Age and condition of the water heater
- Age and condition of the refrigerator and other major appliances

2.3.0 Tightness of the Building Shell

Energy used to heat and cool a home will be wasted if the building shell cannot keep cold out of the home in the winter and heat out of the home in the summer. For that reason, one of the most important parts of the energy audit is checking the building's shell before and after weatherization to determine how tight it is. Ways of checking the tightness of the building shell include the following:

- Visually inspecting the home for obvious defects
- Performing a blower door test on a home to determine the location and extent of air leaks
- Using thermal imagery to identify areas of poor or missing insulation

After the home is weatherized, a follow-up inspection must be done to verify that the weatherization was done properly. This may involve doing another blower door test and using an infrared camera.

2.4.0 Home Lighting

Lighting can consume a great deal of energy. By inspecting the type of lighting used and patterns of lighting usage, the auditor can suggest ways to reduce lighting costs. This can be done by replacing incandescent light bulbs that are used two or more hours a day with more energy-efficient **compact fluorescent lamps (CFLs)** and **light-emitting diodes (LEDs)** and noting lighting usage patterns and suggesting ways they can be modified.

2.5.0 Heat Loss and Heat Gain

To better understand how and where to weatherize a home to prevent energy losses, it is important to understand how a building gains and loses heat energy. It is a law of physics that heat always flows from hot to the cold until it reaches **equilibrium**. Heat will flow faster from hot to cold if the temperature difference between the two is greater. Consider a home that does not have a furnace or air conditioner (*Figure 1*). On a 72°F day, that home's indoor temperature would gradually stabilize until it was the same temperature as the outdoors. If the outdoor temperature went up or down, the home's indoor temperature would track the outdoor temperature.

Now put a furnace and air conditioner in the same home. On a winter day, the home's indoor temperature is 72°F and the outdoor temperature is 30°F (*Figure 2*). In this case, heat would travel from the warmer indoors to the cooler outdoors, causing the home to lose heat. If the furnace did not run, the indoor temperature would slowly drop to the point where it matched the outdoor temperature. When the furnace does run, it replaces any heat that was lost to maintain the indoor temperature at 72°F.

On a summer day, the home's indoor temperature is 72°F and the outdoor temperature is 105°F (*Figure 3*). In this case, heat would travel from the warmer outdoors to the cooler indoors, causing the home to gain heat. If the air conditioner did not run, the indoor temperature would slowly rise to the point where it matched the outdoor temperature. When the air conditioner does run, it removes any heat that was gained to maintain the indoor temperature at 72°F.

The amount of heat that a home loses over a period of time is known as its **heat loss**. The amount of heat a home gains over a period of time is known as its **heat gain**. When a home is being designed, the builder can calculate the exact amount of heat a home will gain or lose. Local climatic conditions, as well as home construction factors, are used in making the calculation. The final number is expressed as **British thermal units** per hour (Btu/h) and is used to size the heating and/or cooling equipment. For example, if a home has a calculated heat loss of 40,000 Btu/h, a furnace capable of producing at least 40,000 Btu/h would be required to heat the home. Factors in home construction that affect heat loss and heat gain include the following:

- Insulation
- Type of windows and doors
- Roofing material
- Tightness of construction

Newer homes that are better insulated, have good quality windows and doors, have the correct roofing material, and are built to modern codes will have lower heat losses and heat gains than older homes. However, older homes can be weatherized so that they gain or lose less heat. That can make them as energy-efficient as newer homes.

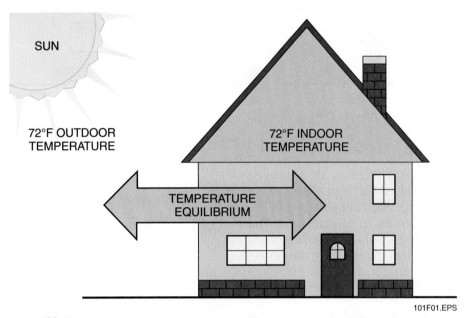

Figure 1 Temperature equilibrium.

Figure 2 Heat loss.

Module 59101-10 Introduction to Weatherization 1.3

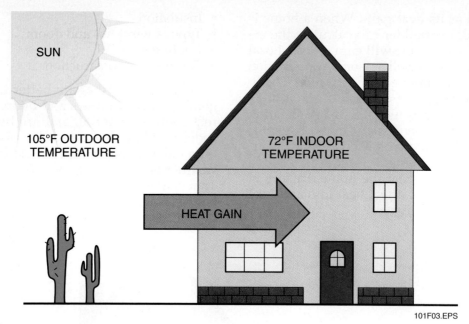

Figure 3 Heat gain.

2.6.0 Air Infiltration

Heat can also be lost or gained as air enters or leaves the home's outer shell through cracks, crevices, and holes (*Figure 4*). Air that enters a home through leaks is known as infiltration. In the winter, cold air entering the house adds to heat loss. In the summer, warm air entering the house adds to heat gain. When conditioned air leaks from the house (exfiltration), it directly affects the utility bill.

The stack effect is a condition present in all buildings that can contribute to air leakage. The stack effect (*Figure 5*) is like what occurs inside a chimney. Warmth at the base of the chimney causes warm air to rise up and out the chimney. As this warm air leaves the home through the chimney, air must be brought into the home to replace it. A home can also act like a chimney. Warm air at lower levels moves to the upper levels. Any openings in the upper level of a home such as an attic door, openings for electrical boxes and ceiling light fixtures, and openings for plumbing pipes can allow air to escape into the attic and out of the home.

The home must replace conditioned air lost through the stack effect. If the home has any openings to the outdoors such as cracks around windows or doors, that is where the outdoor air will enter. Conditioned air can also leave a home through bathroom or kitchen exhaust fans, through a fireplace with an open damper, or through a room air conditioner that is not covered.

A leaking forced-air duct system can also cause a home to lose conditioned air and increase infiltration. In an ideal duct system (*Figure 6*) both the supply air duct and return air duct are leak-free. During operation, the fan in the furnace causes a pressure difference between the supply air duct and the return air duct. High pressure in the supply duct causes the conditioned air to flow into the living space. Low pressure in the return duct causes air in the living space to be drawn back into the return duct. The pressure within the home is neutral under these ideal conditions. This condition causes air to circulate inside the home with little or no losses.

Assume the return duct has no leaks and the supply duct is leaking into an unconditioned space such as a basement, attic, or crawlspace (*Figure 7*). Under these conditions, a slight negative pressure occurs in the living space when the

---

On-Site

Heat Loss and Heat Gain

Several factors including insulation thickness affect the calculation of heat loss and heat gain. There are some differences when calculating the two. The way the building faces the sun affects heat gain, but has no affect on heat loss. Roofs and walls that are shaded from the sun affect heat gain but have no effect on heat loss. Heat loss and heat gain were once calculated using tables and paper forms. Today, computer software is widely used to calculate heat loss and heat gain.

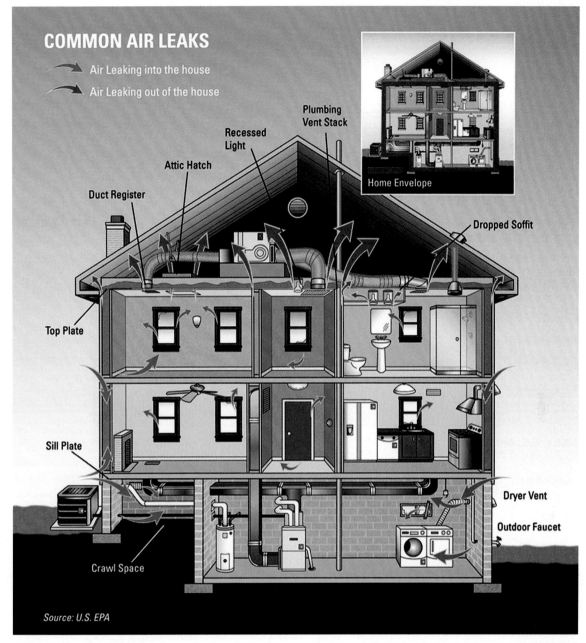

Figure 4 Infiltration and air leaks.

furnace fan runs. The air lost through leaks in the unconditioned space causes this slight negative pressure. To make up for this loss, outside air infiltrates into the home through small openings to the outside. Not only does this waste energy, it causes outdoor air to be drawn into the home.

Assume the supply duct has no leaks and the return duct is leaking in an unconditioned space such as a basement, attic or crawlspace (*Figure 8*). Under these conditions, a slight positive pressure occurs in the living space when the furnace fan runs. This is because the amount of air delivered by the supply duct is greater than the amount of air being drawn from the living space by the return duct. The increased amount of air delivered by the supply duct is a result of outside air that entered the return duct through leaks. Leaks in the return duct can bring in air from both inside and outside the home.

Most forced-air duct systems leak. In a few cases, leaks in the supply and return ducts cancel each other out. In most cases, one duct system will leak more than the other, putting the living space under either a negative or a positive pressure. In either case, outside air can be brought into the home. Leaking ducts can also cause a safety

Module 59101-10 Introduction to Weatherization 1.5

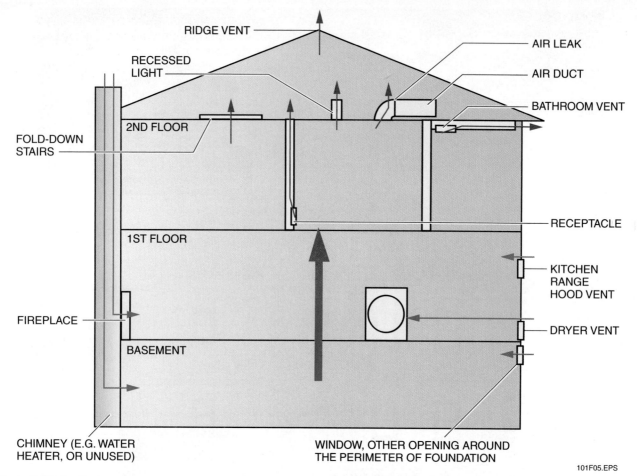

Figure 5 Stack effect.

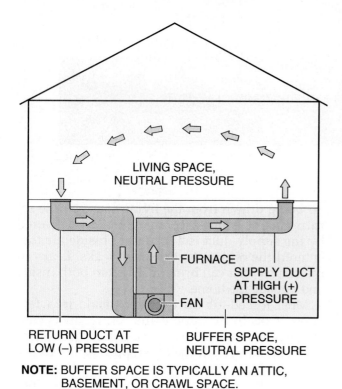

Figure 6 Simplified ideal air duct system.

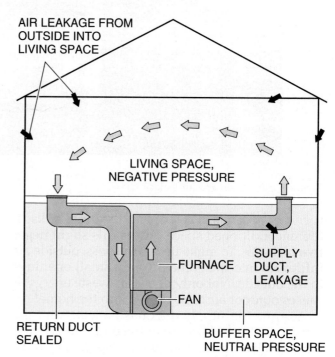

Figure 7 Simplified ideal air duct system with leaks in the supply duct.

1.6 FUNDAMENTALS OF WEATHERIZATION

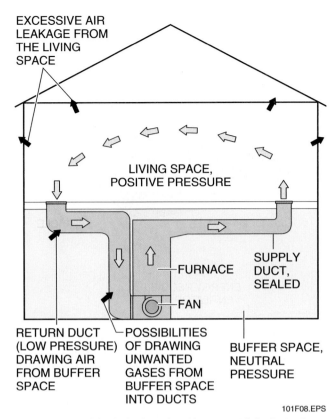

Figure 8 Simplified ideal air duct system with leaks in the return duct.

hazard. For example, a leaking return duct could draw toxic carbon monoxide from a faulty furnace into the duct and then into the home. Leaking return ducts in a basement could also drawn in cancer-causing radon gas.

It would seem that the entry of outdoor air into a home is a bad thing. However, all homes need outdoor air to provide combustion air for fuel burning appliances and to refresh stale indoor air. It is the uncontrolled entry of outdoor air that is inefficient. Outdoor air (ventilation) can be brought into the home under controlled conditions.

3.0.0 Finding Air Leaks and Inadequate Insulation

Before action can be taken to stop air infiltration or to reduce heat loss or heat gain, air leaks in the home have to be found. Places where heat is being lost or gained because they lack insulation must also be identified. These problem areas can be found using one or all of the following methods:

- An inspection of the home
- A blower door to locate air leaks
- An infrared camera to identify areas where insulation is missing

---

On-Site

Balloon Frame Construction

The balloon frame method was once widely used to construct two-story homes. In this method, the outside wall studs continue uninterrupted through the first and second floors. Floor joists on the second floor are attached to the inside edge of the outside wall studs. This often results in a stud cavity that extends from the basement into the attic. These cavities are rarely insulated. An extended stud cavity can contribute to the stack effect and it can also help a fire to spread between floors. For those reasons, balloon frame construction is rarely used today.

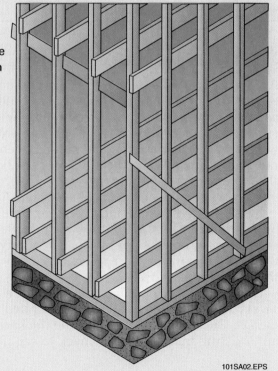

Going Green

Recovery Ventilators

Fresh outdoor air can be brought into a home under controlled conditions using a device called an energy-recovery ventilator (ERV) or heat-recovery ventilator (HRV). These devices use heated or cooled stale exhaust air to heat or cool the incoming air. This results in very little energy being lost in the exhaust air. The device is installed with the heating and/or cooling equipment and operates automatically. It should only be installed in homes that have been properly sealed to prevent air infiltration through openings in the building shell.

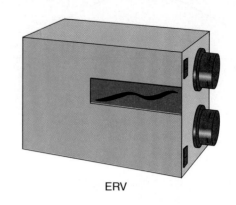

ERV

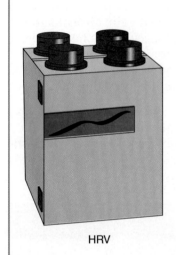

HRV

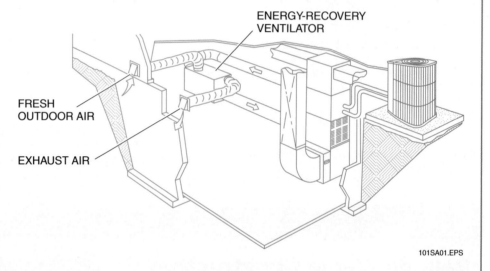

101SA01.EPS

3.1.0 Visual Inspection of the Home

A visual inspection of the home can locate the easy-to-find and obvious problems. It should not be used as the primary means for locating air leaks or poor insulation. Look for gaps or cracks around windows, doors, masonry, and siding where air can enter (*Figure 9*). On the outside of the home, check gaps or spaces where gas, water, or electrical pipes or wires enter or exit the home (*Figure 10*).

Inside the home, check for any ceiling-mounted lights or fans (*Figure 11*). If they are mounted so that they protrude into the attic or cavity between floors, they can cause air to leak into the attic through the stack effect.

If attic access is possible, check that the attic access door is properly sealed and insulated (*Figure 12*). Once in the attic, check the thickness of any existing insulation and check to see if it is installed properly. Note if the attic is properly vented. Look for air leaks and bypasses.

> **WARNING**
> Vermiculite is a mineral that was once widely used as pour-in attic insulation. The product looks like small light-brown or gold pebbles. During mining, some of this product was contaminated with asbestos. If you find vermiculite attic insulation, do not disturb it.

Once in the attic, check the attic floor to see if there are openings to lower floors that might allow air to enter the attic through the stack effect. *Figure 13* shows examples of such openings.

A visual inspection should never take the place of a blower door test for determining air leakage. In fact, a blower door test is a mandatory part of an energy audit.

1.8 FUNDAMENTALS OF WEATHERIZATION

3.2.0 Finding Air Leaks

Air leaks in a home can be found by using a blower door. It can find both obvious and hard-to-find air leaks. The energy auditor performs the pre- and post-blower door test and other tests. Work orders for the weatherization crew are produced as a result of the complete energy audit. Another blower door test is done by the energy auditor after weatherization to verify the effectiveness of the work that was done.

During the weatherization, the crew chief operates the blower door to help pinpoint air leaks and to check the effectiveness of the weatherization as it is being done.

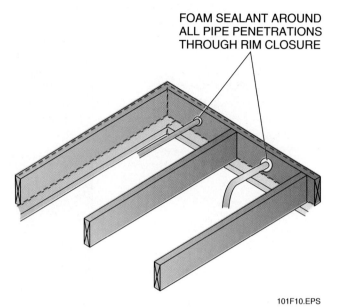

Figure 10 Air leaks around pipes and wires.

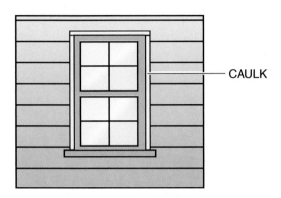

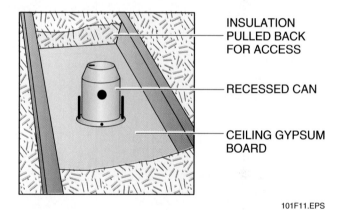

Figure 11 Air leaks around a can light.

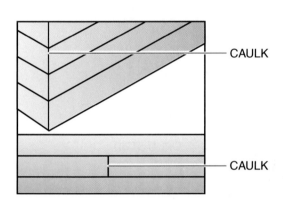

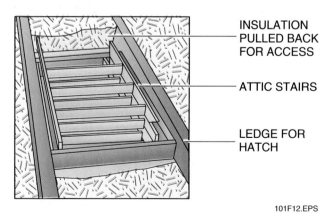

Figure 12 Air leaks around an attic access door.

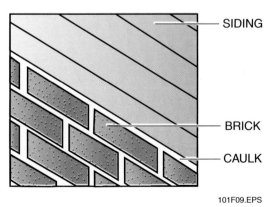

Figure 9 Air leaks through cracks or open seams.

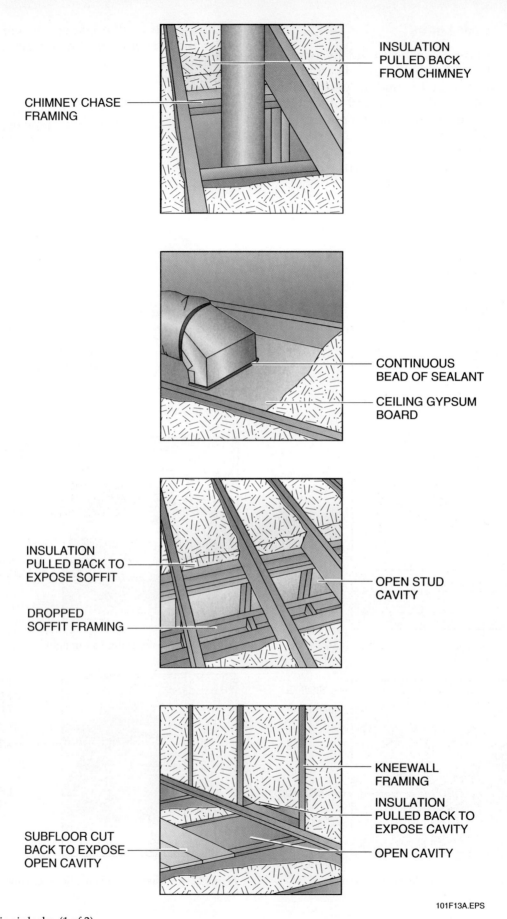

Figure 13 Attic air leaks. (1 of 2)

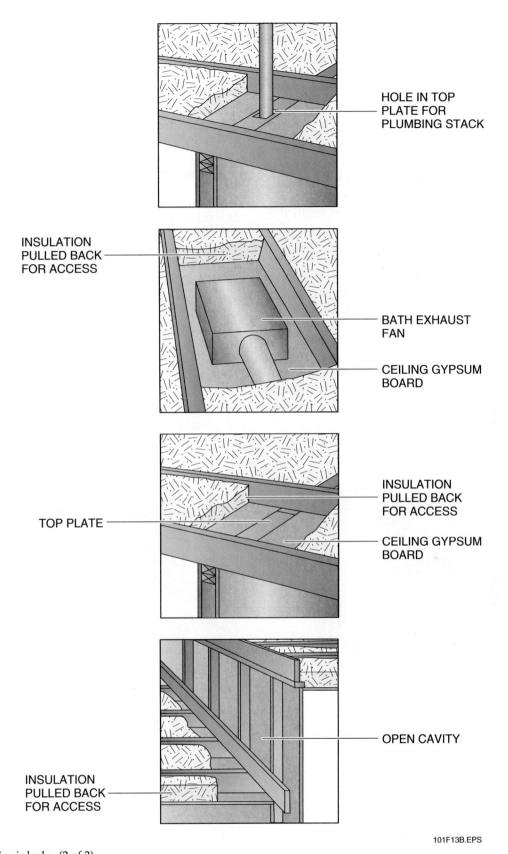

Figure 13 Attic air leaks. (2 of 2)

3.2.1 Finding Air Leaks With a Blower Door

A blower door system consists of a variable-speed fan attached to a frame that can be adjusted to fit different sized doors (*Figure 14*). Instruments to measure pressure differences and airflow (**cubic feet per minute [cfm]**) are part of the system. The blower door can be set up to blow air into (pressurize) the home or to remove air from (depressurize) the home. Depending on how it is set up, the system can cause air to leak into or out of the home.

The most common test method is to set the blower door up so that air is removed from the home. All windows and outside doors must be closed when performing a blower door test. Exhaust fans and the furnace or air conditioner must be turned off so they do not affect the air pressure inside the home. The fireplace damper must also be closed or the fireplace sealed. These steps ensure that the home is sealed so that any air that enters during the test comes in through leaks.

The blower removes air from the home, causing a negative pressure in the home. With the home under negative pressure, air will enter the home through any openings to try to balance pressure with the outdoors (*Figure 15*).

Leaks can be found by going inside the home and feeling near doors, windows, electrical outlets, and light fixtures for air movement. A smoking incense stick, a piece of thread, or a light piece of ribbon (*Figure 16*) can help detect air movement. Large air leaks can sometimes be heard. Instruments on the blower door can determine the leakage rate. Once air leaks are found, they can be sealed.

3.2.2 Finding Inadequate Insulation

Excessive heat loss and/or heat gain can often be traced to poor or missing insulation. Visual inspection of insulation in the basement, attic, and crawlspace can easily verify if the insulation there is adequate. But how can insulation hidden behind walls or ceilings be checked for adequacy? An infrared camera (*Figure 17*) is often used to check how well insulation is doing its job.

The infrared camera is aimed at the section of the home in question. The color image on the camera's screen will tell if heat is being lost or gained. Heat is indicated by red or orange. A lack of heat is indicated by cooler colors such as blue or green. *Figure 18A* shows a conventional view of the outside of a home and *Figure 18B* shows an infrared view

INTERIOR

EXTERIOR

Figure 14 Blower door installed in an exterior door.

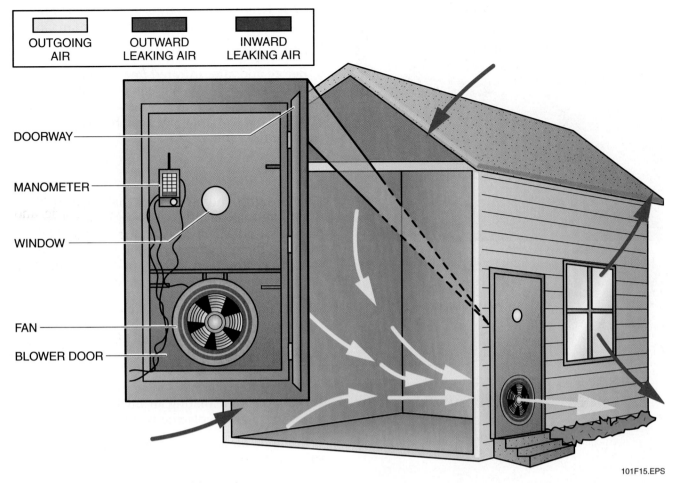

Figure 15 Finding air leaks with a blower door.

Figure 16 Finding an air leak.

Figure 17 Infrared camera.

Module 59101-10 Introduction to Weatherization 1.13

Figure 18 Conventional and infrared images.

of the outside of the same home. In the winter, the infrared view of the outside of a home should not show any red or orange. If red or orange is seen, it shows that heat is leaving the home.

The infrared camera can be used inside or outside of the home. In the summer, it can be used inside to see if heat is entering the home. For example, if the infrared camera shows a red or orange spot on the ceiling, it means that heat is entering the home through the ceiling. The infrared camera is useful in helping to find air leaks when used with a blower door. If the home is not pressurized, an air leak may not be visible. To help find an air leak using an infrared camera and a blower door, follow these steps:

Step 1 Record an infrared image of the area in question with the blower door fan OFF.

Step 2 Turn the blower door ON. The home can be under either negative or positive pressure.

Step 3 Record an infrared image of the area in question with the blower door fan ON.

Step 4 Compare the two infrared images.

Before and after infrared images show how the blower door made the air leak visible by drawing in heated air. Without the blower door causing airflow, that leak may not have been apparent to the infrared camera.

4.0.0 Weatherizing a Home

Weatherizing a home involves making changes to the home that result in decreased heat loss and heat gain and a decrease in air infiltration. By doing this, the homeowner gains by having lower costs for heating and cooling. In the U.S., the steps required to weatherize a home vary from region to region. Steps needed to weatherize a home may include the following:

- Sealing the home to prevent air from leaking in or out
- Adding insulation
- Making the roof more energy-efficient
- Sealing and insulating air ducts
- Upgrading or replacing windows and doors

Methods for sealing air leaks and applying insulation to reduce heat loss and heat gain will be taught in *Sealing the Building Envelope*.

4.1.0 Adding Insulation

Insulation acts to slow down the flow of heat from hot to cold. In the winter, this reduces heat loss and slows down the flow of heat out of the home. With less heat loss, a smaller furnace can be used to heat the home. In the summer, insulation reduces heat gain and slows down the flow of heat into the home. With less heat gain, a smaller air conditioner can be used to cool the home.

4.1.1 Types of Insulation

A number of different types of insulation are available to insulate a home (*Figure 19*). How well insulation slows down the flow of heat is found in its **R-value**. The R-value is stated as a number,

---

On-Site

Blower Door Versatility

With the proper accessories, a blower door can be used to detect air duct leakage, as well as leakage in the building envelope.

such as R-13 or R-49. A higher R-value means the material is a better insulator. Common types of insulation include the following:

- Flexible insulation
- Rigid foam board
- Loose-fill insulation
- Spray-in-place insulation
- Spray foam insulation

4.1.2 Flexible Insulation

Flexible insulation (*Figure 20*) is widely used to insulate walls, floors, and ceilings. It is available in batts in various widths and R-values and can come with or without a vapor barrier. Thicker insulation has a higher R-value. When weatherizing a home, flexible insulation without a vapor barrier is often laid on top of existing attic insulation. The R-value of the new insulation adds to the R-value of any existing insulation. The total R-value of attic insulation will depend on local climatic conditions, and local codes.

4.1.3 Rigid Foam Board

Rigid or semi-rigid foam boards (*Figure 21*) are available in sheet or board form in sizes up to 4 feet wide and 12 feet long. Sheets can be from ⅜ inch to 4 inches thick with R-values up to R-25. The boards are formed of fiberglass or foamed plastic and are water-resistant. Foam boards can be used to insulate basement walls, or they can be used where a rigid structure is needed, such as a closure board (*Figure 22*). Local fire codes typically regulate the use of rigid foam board.

> ### On-Site
> ### Insulation and Air Leaks
> Many people think that if an area is insulated, the insulation will stop air leaks. Here's why that is not always true. Insulation will slow down the flow of heat. If the insulation is foam board or other solid material, it will also stop airflow. But what if flexible insulation without a vapor barrier is used? Flexible insulation will allow air to pass right through unless some kind of air barrier is used with it.

BLANKET INSULATION

BATT INSULATION

Figure 19 Different types of insulation.

Figure 20 Flexible insulation.

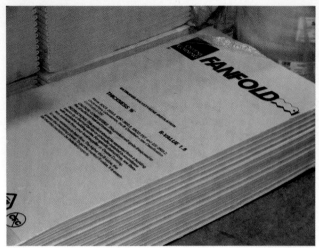

Figure 21 Rigid foam board.

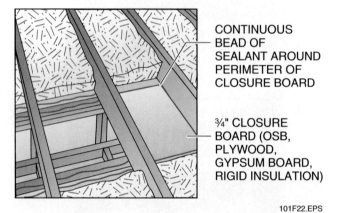

Figure 22 Closure board made of rigid foam board.

4.1.4 Loose-Fill Insulation

Loose-fill insulation can be made of many different materials including fiberglass, shredded bark, and ground-up newspapers. It is shipped in bags and can be blown over existing attic insulation (*Figure 23*). It can also be blown into wall cavities of homes that do not have any wall insulation. The R-value of loose-fill insulation is given as a per-inch value. If an R-2 per inch value were given, 10 inches of the insulation would yield an R-value of R-20.

4.1.5 Spray-in-Place Insulation

Spray-in-place insulation is a fibrous material that is either treated with an adhesive or is sprayed against an adhesive surface (*Figure 24*). This material is more likely to be used in new construction for windows and doors. A trained operator using special equipment must apply spray-in-place insulation.

GOING GREEN

Recycled Material

Shredded bark and ground-up newspapers used to make loose-fill insulation are materials that might otherwise be thrown away. Recycling such material into a useful product helps protect the environment.

Figure 23 Applying loose-fill insulation.

Figure 24 Applying spray-in-place insulation.

4.1.6 Spray Foam Insulation

Spray foam insulation is a chemical that is sprayed on and sets up to form rigid insulation. Foam insulation is available in single spray cans or portable two-part insulation kits (*Figure 25*). Their small size makes them ideal for weatherization work. Foam is often used to seal windows and doors as well as openings where pipes and wires pass through walls or ceilings (*Figure 26*).

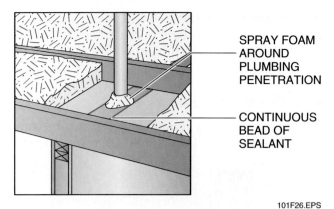

Figure 26 Sealing an air leak with spray foam.

4.2.0 Sealing Air Leaks

There are a number of places on the outside of a home where air can leak into or out of the structure. These include open seams where siding butts up to a window or door, around **rim joists**, and openings where pipes or wires enter or exit the building. Doors and windows that do not seal properly also leak air. The individual leaks are usually small. When these small leaks are added up, however, they represent a major air leak. Simple steps such as caulking and sealing open joints, and adding weatherstripping to doors and windows

Figure 25 Portable spray foam kits.

can greatly reduce these losses. Weatherization technicians will learn how to caulk and seal leaks and install weatherstripping in *Sealing the Building Envelope*.

4.2.1 Caulks and Sealants

Caulk is available in rope form, or in a tube that is applied with a caulking gun (*Figure 27*). Choose a caulk that will remain flexible and that can be painted. Rope caulk is stiffer and easier to handle than caulk from a tube. It can fill wider cracks and can be molded around pipes and odd shapes. Spray foam insulation is also useful for sealing larger openings. It is commonly used to seal around windows and doors.

4.2.2 Weatherstripping

Loose fitting windows and doors are common sites of air leakage. To stop this leakage, weatherstripping can be applied around the edges of the door or window to stop air from entering. To seal double-hung windows, use tubular or strip materials (*Figure 28*). Sliding windows are sealed in a similar manner (*Figure 29*). Seals for the bottom of a door can be made of vinyl, rubber, or metal (*Figure 30*).

4.3.0 Losses Through Windows and Doors

Glass is a good conductor of heat. For that reason, a lot of heat is lost or gained through window glass. As much as 38 percent of the cooling load and up to 20 percent of the heating load can be tied to losses through glass (*Figure 31*).

Figure 27 Apply caulk with a caulking gun.

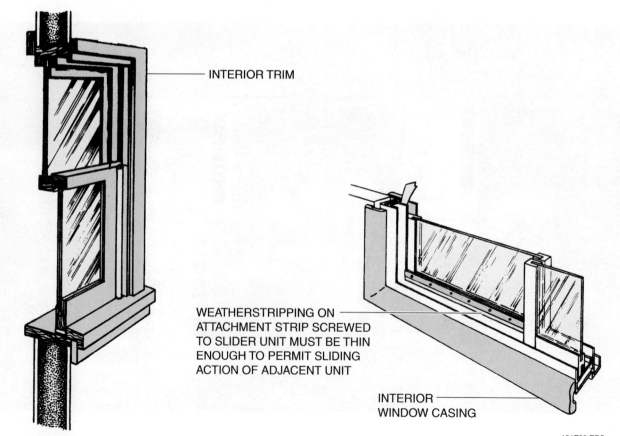

Figure 28 Sealing air leaks around a double-hung window.

On-Site

Energy-Efficient Glass

Low-E glass is coated with a thin metallic substance that makes it very good at controlling radiant heat transfer. Special heat-absorbing glass is also available that contain tints that absorb about 45 percent of incoming heat from the sun. The sun's heat is transferred from the window to the home. If low-E glass cannot be used, a reflective film can be applied to existing windows. This film helps stop heat loss in the winter and reflects sunlight in the summer.

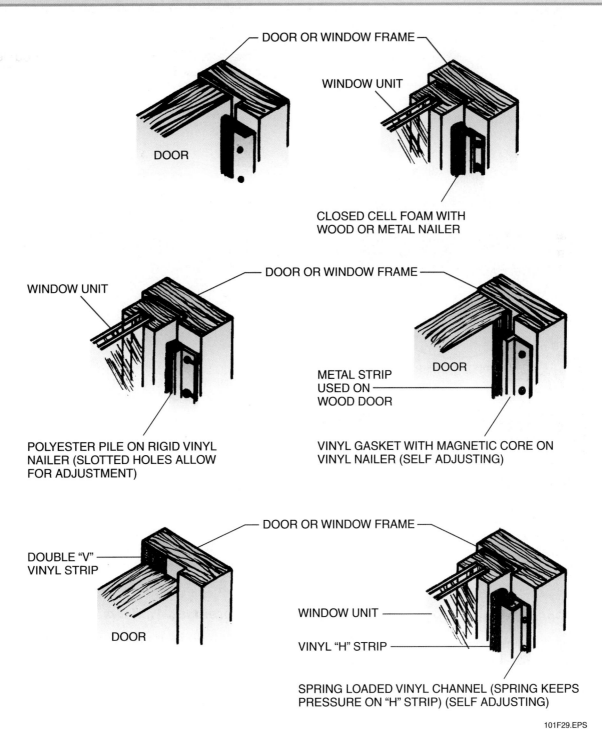

Figure 29 Sealing air leaks around sliding windows.

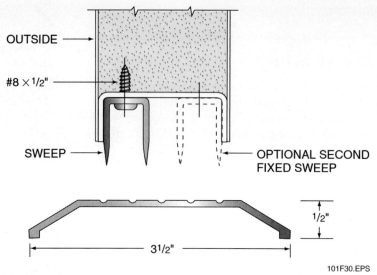

Figure 30 Sealing air leaks at the bottom of a door.

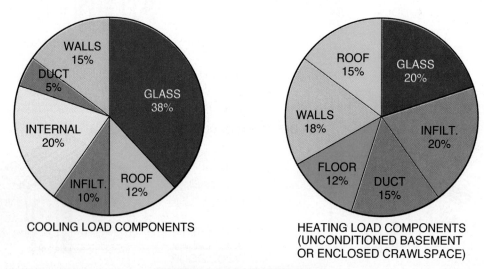

Figure 31 Effect of glass on heating and cooling loads.

The number of panes of glass in a window and the spacing between the panes determines the **U-value** of the glass (*Table 1*). The lower the U-value, the greater the window's resistance to heat flow. Single pane glass has the greatest heat loss.

If an outside door contains a lot of glass, it can be treated as a window for heat loss and heat gain purposes.

4.3.1 Upgrading Windows and Doors

Windows and doors may be upgraded or replaced as part of a weatherization program if they cannot be repaired. However, replacements are not given a high priority in funded programs because the efficiency gain is not as great as that of other improvements. Replacing all windows and doors can be costly. If cost is a problem, existing windows and doors can be upgraded to improve their efficiency. Examples of upgrades include the following:

- Replacing cracked or broken glass
- Caulking around window or door frames
- Adding weatherstripping
- Repairing or replacing storm windows and storm doors

4.3.2 Replacement Windows

If windows are to be replaced as part of a home weatherization, choose a window with double- or triple-pane glass (*Figure 32*). If the budget allows, consider installing ultra-efficient **low-E glass**. These windows significantly reduce radiant heat transfer into the building.

Table 1 U-Values of Window Glass

Window Type	U-Value
Single pane	1.099
Single pane with storm	0.500
Double pane (3/16" air space)	0.620
Double pane (3/4" air space)	0.420
Double pane (1/2" airspace with low-E)	0.320
Double pane (with suspended film and low-E)	0.240
Triple pane (1/2" air space)	0.310

4.3.3 Replacement Doors

Replacement doors can be made of wood, metal, or fiberglass. Most are insulated and have a U-value. When choosing a replacement door, select one that stays within the budget and is energy-efficient.

4.4.0 Energy-Efficient Roofs

The color of a roof can have a direct bearing on home energy use. A dark roof can absorb a lot of solar heat. That heat is then radiated from the underside of the roof into the attic where it increases the cooling load. This problem is greatest in warm climates. Replace dark roof shingles with lighter-colored roof shingles to reduce heat gain. Built-up roofs can be coated with a white or aluminized coating that reflects the sun's rays.

4.5.0 Sealing and Insulating Air Ducts

Ductwork in older homes is a common source of lost heating efficiency. Over time, material used to seal duct connections can deteriorate, resulting in air leakage.

If the ductwork leaks, one of two things can happen. In the supply air system, heated air can

> **Going Green**
>
> **Trees**
>
> Trees placed on the south- and west-facing sides of a home can shade the walls and roof, reducing summer heat gain. As an added bonus, trees remove carbon dioxide from the atmosphere. Carbon dioxide is a greenhouse gas that contributes to global warming.

leak out of the ductwork and fail to reach the areas that need heat. In a return air system, cold air can be drawn into the ductwork, reducing the efficiency of the heating system and altering the pressure balance of the home.

Air ducts must be sealed to prevent conditioned air from being lost. A special type of mastic is generally used for this purpose (*Figure 33*). Adding insulation to sheet metal ducts is another way to prevent these ducts from gaining or losing heat. Both of these steps save energy. You will learn how to seal and insulate air ducts in *Insulating Pipes, Ducts, and Water Heaters*.

5.0.0 REDUCING THE BASELOAD

Home energy use falls into two categories: seasonal use and year-round use. Seasonal use is the energy used to heat and cool the home. It tends to peak during the coldest part of the winter and the hottest part of the summer (*Figure 34*). Year-round use, on the other hand, remains fairly steady throughout the year. It increases a bit in the winter because the days are shorter, and bad weather keeps people in their homes. Year-round energy use, also called the baseload, can also be reduced

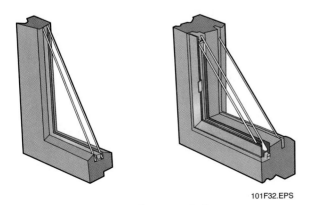

Figure 32 Double- and triple-pane windows.

Figure 33 Duct joints sealed with mastic.

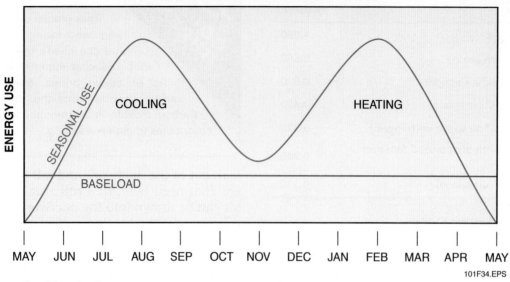

Figure 34 Seasonal and baseload energy use.

to bring down overall energy use. Weatherization reduces heat loss and heat gain and directly impacts seasonal energy use. The energy auditor can advise and educate the homeowner on ways to reduce the baseload.

According to research from Lawrence Berkeley National Laboratory, typical household energy use is broken down as follows:

- *Heating* – 29 percent
- *Cooling* – 17 percent
- *Water heating* – 14 percent
- *Appliances (refrigerator, dishwasher, washer and dryer)* – 13 percent
- *Lighting* – 12 percent
- *Other (coffee makers, battery chargers, ceiling fans, etc.)* – 11 percent
- *Electronics (computer, TV, etc.)* – 4 percent

It is apparent that over half the energy consumed in the home is not used for heating and cooling. Baseload reduction is focused on three areas that consume household energy use: appliances, water heating, and lighting.

Appliances responsible for increased energy use include refrigerators, freezers, washing machines, clothes dryers, and dishwashers. Refrigerators and freezers are the worst offenders. Inefficient washing machines and dishwashers use lots of hot water. The wasteful use of hot water increases the cost to heat water.

5.1.0 Appliances

Older electric appliances, especially refrigerators and freezers, are expensive to operate. Replacing an older appliance with an Energy Star® appliance can greatly reduce baseload energy use. The following are some average electrical power or energy savings that can be had when using Energy Star® appliances:

- *Refrigerator* – Up to $80/year
- *Dishwasher* – Up to $30/year
- *Washing machine* – Up to $145/year
- *Water heater* – Up to 50 percent energy reduction
- *Coffee maker* – Up to $80/year

Many electrical devices in modern homes steal energy when not in use. For example, when a computer or flat-screen TV is turned off, it is often actually in a standby mode where it continues to use power. These types of devices are often called parasitic loads.

5.1.1 Refrigerators

Older refrigerators and freezers built before 1990 can use up to three times the power of a newer, more energy-efficient Energy Star® model. The energy auditor can recommend replacing the old refrigerator. If that is not an option, the auditor can recommend steps than can reduce refrigerator power use, including the following:

- Cleaning the condenser coil
- Replacing leaking door seals
- Adjusting the thermostats so that the freezer temperature is set between 0°F and 5°F and the refrigerator temperature is set between 36°F and 40°F
- Changes in use such as allowing cooked foods to cool to room temperature before refrigerating

On-Site

Radiant Barriers

In desert areas of the U.S. that have very hot summers, radiant barriers are sometimes installed in attics. They prevent the heat that radiates from the underside of the roof from entering the home. The barrier can be suspended below the roof rafters or the barrier can take the form of thin metallic-faced plastic chips that are blown in on top of the insulation.

5.1.2 Other Appliances

Other major household appliances contribute to baseload energy use as well. Replacing them with more efficient models reduces the baseload. Improved technology has made dishwashers more energy-efficient. Energy Star® dishwashers can save at least $30 per year over models made before 1994. Newer models use less hot water than older models, which helps reduce the cost of heating water.

Newer washing machines achieve lower operating costs because they use less hot water and they spin clothes to remove more water. Clothes that are not as wet dry faster in a dryer. Replacing a washing machine that is over 10 years old with an Energy Star® model can reduce energy costs about $145 per year.

Clothes dryers can be costly to operate. An electric dryer costs about one cent per minute to operate. It is easy to see that the cost to dry the clothes of a large family could get very expensive over time. Energy Star® does not rate clothes dryers.

5.1.3 Water Heaters

Energy to heat water contributes to baseload energy use. Appliances that waste hot water can add to those losses. The water heater can lose heat through its outer casing and through the pipes that carry the heated water throughout the home. Wrapping the water heater in an insulating jacket and insulating hot water pipes can reduce these losses (*Figure 35*). Weatherization technicians will learn how to insulate a water heater and water pipes in *Insulating Pipes, Ducts, and Water Heaters*.

If the water heater is leaking or defective, replace it with an Energy Star® water heater. Installing a tankless or on-demand water heater can further reduce water-heating costs. This type of water heater only heats water when it is needed. This eliminates the need to maintain a tank full of hot water.

5.2.0 Lighting

Lighting is a big contributor to baseload energy use. There are two areas in which lighting energy use can be reduced: replacing inefficient lighting and modifying lighting usage.

Incandescent light bulbs are energy hogs. Simply replacing all incandescent bulbs in a home with CFLs or LEDs (*Figure 36*) can greatly reduce energy use. For example, a compact fluorescent lamp that is the equivalent of a 60-watt incandescent bulb only consumes 13 watts of power. While these newer lamps cost more, payback is obtained through lower operating costs and longer lamp life. LEDs are another type of energy-efficient source of light. This new technology is fairly expensive, but the cost is expected to come down as they become more available.

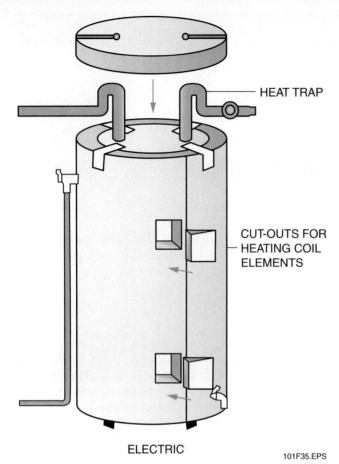

Figure 35 Water heater wrapped to prevent heat loss.

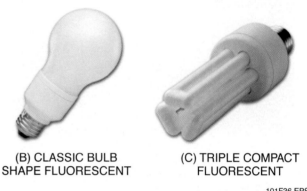

Figure 36 Compact fluorescent lamps.

5.3.0 The Energy Auditor as an Educator

The energy auditor can help further reduce household energy use by educating homeowners. The auditor can make the homeowner aware of the energy savings that can be realized by changing use habits and making simple and inexpensive improvements to the home. Changes in habits include the following:

- Washing only full loads of clothes
- Washing clothes in cold water
- Drying only full loads of clothes
- Washing only full loads of dishes
- Setting the room thermostat to a lower level for heating and a higher level for cooling
- Shutting off heat in unoccupied rooms
- Taking shorter showers
- Turning off TVs and personal computers if not in use
- Turning off lights in unoccupied rooms
- Closing drapes to keep heat out in summer and in during winter

Examples of inexpensive home improvements that can result in energy savings include the following:

- Installing light sensors to control outdoor lighting
- Installing motion sensors to turn on security lighting
- Installing a programmable room thermostat
- Installing a water-saving shower head
- Installing solar screens on windows
- Installing awnings for window shade
- Installing thermal window drapes
- Planting trees for summer shade
- Putting up a clothesline

GOING GREEN

Refrigerants

Older refrigerators and freezers are likely to use a refrigerant that has been shown to damage the ozone layer that surrounds the planet. Removing an old refrigerator from service also permanently removes that refrigerant. However, by law the refrigerant must be recovered and disposed of properly.

Going Green

Energy Star®

Household appliances made for sale in the U.S. may earn an Energy Star® rating. Energy Star® is a program of the U.S. Environmental Protection Agency (EPA). Products that meet Energy Star® requirements are energy-efficient. Use of a rated appliance will result in energy savings for the homeowner as well as reduced air pollution due to more efficient use of energy. For more information about the Energy Star® program, go to www.energystar.gov.

On-Site

Checking Appliance Power Use

Portable power meters are available that allow energy auditors to check how much power an electric appliance is using. The auditor can then compare current power use with the reduced power use of an Energy Star appliance.

6.0.0 CAREERS IN WEATHERIZATION

The push to weatherize homes to make them more energy-efficient, and the availability of funding to make it happen, will create many new jobs in this field. Each of these jobs requires a different set of skills. Types of jobs that will be created include weatherization technician, weatherization crew chief, and energy auditor.

6.1.0 Weatherization Technician

The weatherization technician is the person who will perform the tasks required to weatherize a home. He or she must have the manual skills needed to find and seal air leaks, add insulation, and perform other manual tasks. In addition to manual skills, the technician must be able to work with other members of the crew as part of a team and must also be able to communicate with homeowners and other job-related contacts. This is an entry-level position that can lead to advancement as additional skills and experience are gained.

6.2.0 Weatherization Crew Chief

The weatherization crew chief supervises a crew of weatherization technicians, and checks their work, including the blower door test. The crew chief must have all the manual skills required of a weatherization technician, as well as supervisory skills. He or she must be able to schedule work and material flow, communicate with homeowners, and deal with worker issues such as tardiness or absenteeism. Crew chiefs must also work with and satisfy the needs of their own supervisors. A crew chief can be a person who started as a weatherization technician and was promoted based on job performance and additional training.

6.3.0 Energy Auditor

The energy auditor does the initial survey of a home to determine what needs to be done to make it more energy-efficient. He or she must be trained to use test equipment such as a blower door and an infrared camera to evaluate the home. The auditor will prepare a detailed report and/or work order showing what must be done during the weatherization, and will then conduct a follow-up visit to ensure that the job was properly done. The energy auditor must have good communications skills to effectively deal with customers, the weatherization crew, and upper management. The energy auditor must be properly trained to perform this job. A person experienced as a weatherization technician or crew chief could qualify for promotion to energy auditor by completing additional training.

6.4.0 Standardized Training by NCCER

The National Center for Construction Education and Research (NCCER) is a not-for-profit education foundation established by the nation's leading construction companies. NCCER was created to provide the industry with standardized construction education materials, the Contren® Learning Series, and a system for tracking and recognizing students' training accomplishments—NCCER's National Registry. Refer to the *Appendix* for examples of NCCER credentials.

NCCER also offers accreditation, instructor certification, and skills assessments. NCCER is committed to developing and maintaining a training process that is internationally recognized, standardized, portable, and competency-based.

Working in partnership with industry and academia, NCCER has developed a system for program accreditation that is similar to those found in institutions of higher learning. NCCER's ac-

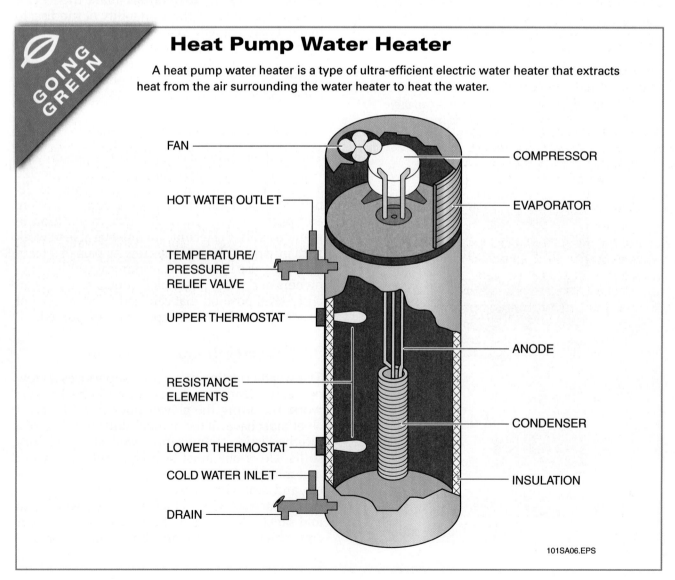

Heat Pump Water Heater

A heat pump water heater is a type of ultra-efficient electric water heater that extracts heat from the air surrounding the water heater to heat the water.

1.26 FUNDAMENTALS OF WEATHERIZATION

Going Green: Programmable Room Thermostats

Installing a programmable room thermostat can reduce seasonal energy use. The device can be set up to adjust the home's temperature up or down based on occupancy and/or the time of day. For example, if there is no one home during the day, the heat setting can be lowered or the cool setting raised. The thermostat can then automatically raise or lower the temperature setting so the home is comfortable when the occupants are home.

101SA07.EPS

creditation process ensures that students receive quality training based on uniform standards and criteria. These standards are outlined in NCCER's Accreditation Guidelines and must be adhered to by NCCER Accredited Training Sponsors.

More than 450 training sponsors and/or assessment centers across the U.S. and eight other countries are proud to be NCCER Accredited Training Sponsors and Accredited Assessment Centers. Millions of craft professionals and construction managers have received quality construction education through NCCER's network of Accredited Training Sponsors and the thousands of Training Units associated with the Sponsors. Every year the number of NCCER Accredited Training Sponsors increases significantly.

A craft instructor is a journeyman craft professional or career and technical educator trained and certified to teach NCCER's Contren® Learning Series. This network of certified instructors ensures that NCCER training programs will meet the standards of instruction set by the industry. There are more than 4,300 master trainers and 45,000 craft instructors within the NCCER instructor network. More information is available at www.nccer.org.

6.5.0 Advancement Opportunities In Weatherization

The training you receive in the weatherization program should be viewed as the first step on your career ladder in the construction industry. Once you receive your initial training, you can begin to build experience. You will gain experience doing job-specific skills, but you will also gain other skills such as working as part of a team and interacting with people from all walks of life.

Your initial job experience can provide access to increasing opportunities in weatherization, leading to responsible positions and job security.

Summary

The U.S. government has committed billions of dollars to home weatherization. The benefit is lower energy costs, less dependence on foreign energy sources, and cleaner air through reduced carbon dioxide and other greenhouse gas emissions.

Weatherizing a home consists of finding and correcting energy problems in the home that allow heat to be lost in the winter and gained in the summer. Before weatherization can take place, the sources of the losses must be found through an energy audit. Advanced methods such as a blower door test and infrared imaging of the home are used for this purpose.

Energy-wasting heat loss and heat gain can be corrected by sealing air leaks and openings (bypasses) and openings in the building, upgrading or replacing inefficient doors and windows, adding insulation, and changing the color of the roof.

Weatherization helps reduce seasonal energy use. Energy that is used year round, called baseload energy, must also be reduced. Appliances such as refrigerators and dishwashers, water heaters, and lighting consume baseload energy. Replacing old appliances with Energy Star® appliances, replacing incandescent light bulbs with compact fluorescent lamps, and changing energy use habits can reduce the baseload.

Careers in weatherization can take different tracks. Once on the job, weatherization technicians can advance to crew chief, energy auditor, or to other jobs in the construction field. NCCER has developed a competency-based training program to help you achieve your career goals.

Review Questions

1. According to the *ARRA of 2009*, homeowners can receive tax credits for installing a _____.
   a. solar water heater
   b. clothesline
   c. furnace
   d. wind tunnel

2. When it is 65°F outdoors and 65°F inside a home, in what direction will heat flow?
   a. Into the house.
   b. Out of the house.
   c. Out of the house through the attic.
   d. There would be no heat flow.

3. An air conditioner runs in the summer because a home has a heat loss.
   a. True
   b. False

4. The stack effect is similar to what takes place in a(n) _____.
   a. chimney
   b. air duct
   c. wind turbine
   d. greenhouse

5. What happens to the pressure within a home if the supply air duct is leaking into the attic that is outside the envelope?
   a. The pressure remains neutral.
   b. The pressure becomes positive.
   c. The pressure becomes negative.
   d. It causes the furnace to run longer.

6. Under what conditions can outside air be brought into a home?
   a. Under controlled conditions.
   b. Only in summer if the temperature is above 72°F.
   c. Only if the home has a fuel-burning furnace.
   d. Outside air must not be brought into the home.

7. What is often used to find small air leaks around a window or door?
   a. A stethoscope
   b. An incense stick
   c. A microphone
   d. A calibrated flow hood

8. Loose-fill insulation is typically shipped in boxes.
   a. True
   b. False

9. How much of the total cooling load can be assigned to losses through window glass?
   a. Up to 38 percent
   b. Up to 24 percent
   c. Up to 16 percent
   d. Up to 10 percent

10. Which of the following windows would provide maximum energy efficiency?
    a. Single pane of glass
    b. Single pane with storm window
    c. Double-pane glass
    d. Low-E double pane glass

11. An example of an energy-using device that contributes to the baseload is a(n) _____.
    a. furnace
    b. air conditioner
    c. television
    d. heat pump

12. In the typical U.S. household, most of the energy consumed is used for _____.
    a. heating
    b. operating appliances
    c. drying clothes
    d. cooling

Review Questions

13. How much more power will a refrigerator built before 1990 use compared to an Energy Star® model?

    a. Three times more
    b. Two times more
    c. 40 percent more
    d. 25 percent more

14. Energy used for lighting can be reduced by _____.

    a. replacing all fluorescent lamps with incandescent bulbs
    b. converting all lighting fixtures to 240 volts
    c. using motion sensors to control security lighting
    d. using mercury vapor bulbs throughout the home

15. Which of the following is an example of a change of habit than can lead to energy savings in the home?

    a. Install compact fluorescent lamps in all fixtures.
    b. Install a tankless water heater.
    c. Put up a clothesline.
    d. Wash clothes in cold water.

Trade Terms Quiz

Fill in the blank with the correct trade term that you learned from your study of this module.

1. Air within a home that has been heated or cooled is called _____.

2. Year-round energy use is called the _____.

3. The unwanted entry of air into a home is called _____.

4. The _____ indicates the ability of insulation to resist the flow of heat.

5. Use _____ to prevent radiant heat from entering or leaving a home.

6. A(n) _____ contains a variable-speed fan and is used to detect air leaks in a home.

7. How often a furnace runs depends on a home's _____.

8. A U.S. government program that rates products on their energy efficiency is called _____.

9. The _____ brings air to a furnace or air conditioner.

10. A(n) _____ is very energy-efficient and is known for long life.

11. Heated air is carried out through the attic by the _____.

12. The _____ carries heated or cooled air to rooms in a home.

13. To prevent water from condensing on a cold surface, insulation must have a(n) _____.

14. The _____ is a measurement term that defines the quantity of heat that a home loses or gains.

15. Heat that enters a home is called _____.

16. A thermal image of a home is produced by a(n) _____.

17. _____ is a radioactive gas that can cause lung disease.

18. A board that encloses the ends of the floor joists is called a _____.

19. When opposing forces reach a balanced condition, it is known as _____.

20. Airflow is typically measured in _____.

21. _____ occurs when heated or cooled air escapes from the home.

22. An efficient but expensive type of lighting is called a(n) _____.

23. An energy-efficient window will have a low _____.

Trade Terms

Baseload
Blower door
British thermal unit (Btu)
Compact fluorescent lamp (CFL)
Conditioned air
Cubic feet per minute (cfm)
Energy Star®
Equilibrium
Exfiltration
Heat gain
Heat loss
Infiltration
Infrared camera
Light-emitting diode (LED)
Low-E glass
Radon
Return air duct
Rim joist
R-value
Stack effect
Supply air duct
U-value
Vapor barrier

Cornerstone of Craftsmanship

Steve Buglione
Weatherization Manager
McNeal and White Contractors
Jacksonville, FL

Steve Buglione worked his way up from an entry level job as a construction laborer to a construction project manager. Today he is an area manager for weatherization projects.

How did you get started in the construction industry?
I started out at the bottom of the totem pole as a laborer, but through hard work and learning everything I could along the way, I worked my way up to project manager within eight years.

What inspired you to enter the construction industry?
My wife played a major role. She knew that I really liked building, so she encouraged me to enter the field, even though I had to start at the bottom. If you talk to anyone who has had a successful career, you will probably learn that they were inspired and encouraged by a family member, friend, supervisor, or teacher. It's much easier to take the risk of entering a new field when you have the support and encouragement of others. You will also find that they were willing to work hard and learn all they could about their chosen career field.

What do you enjoy most about your job?
I really like construction work, and the satisfaction I get from completing a job. I also get a great deal of satisfaction from improving the lives of others through my work in weatherization and construction. I think about how my work in weatherization helps people. It makes their homes more comfortable and reduces their cost of living. For many people, it can be the factor that allows them to remain in their homes. I really enjoy seeing the look of happiness on a homeowner's face when we present a completed job.

How important do you think NCCER credentials are?
I think those credentials are extremely important. First of all, they allow potential employers to immediately see the knowledge and skills a person has. Also, well-structured, competency-based training is essential in today's job market. The industry has become more technically complex in terms of materials, tools, and equipment. A person who lacks training or fails to keep up with trends in the industry is going to have a difficult time getting and keeping a job and certainly will have trouble in making career headway.

Would you suggest construction as a career to others?
Yes. There are many opportunities in the construction industry. It is a source of great satisfaction because you can see the results of the work you do, often for years to come. A person in this industry can earn an excellent income and has a lot of opportunity for advancement.

How do you define craftsmanship?
I think a craftsman is a person who has made the effort to learn the necessary job skills and takes pride in the work he or she does.

Appendix

SAMPLES OF NCCER TRAINING CREDENTIALS

NATIONAL CENTER FOR CONSTRUCTION EDUCATION AND RESEARCH

August 27, 2009

Sample Student
National Center for Construction Education and Research
3600 NW 43rd St Bldg G
Gainesville, FL 32606

Dear Sample,

On behalf of the National Center for Construction Education and Research, I congratulate you for successfully completing NCCER's Contren® Learning Series program. I also congratulate you for choosing construction as a career.

You are now a valuable member of one of our nation's largest industries. The skills you have acquired will not only enhance your career opportunities, but will help build America.

Enclosed are your credentials from the National Registry. These industry-recognized credentials give you flexibility in planning your career and ensure your achievements follow you wherever you go.

To access your training accomplishments through the Automated National Registry, follow these instructions:
1. Go to www.nccer-anr.org.
2. Click the "Individuals" button.
3. Enter the NCCER card number, located on front of your wallet card or transcript, and your PIN.
    Note: The default PIN is the last four digits of your SSN. You may change your PIN after you login.
4. First-time users will be directed to answer a few security questions upon initial login.
5. Contact the registry department with any questions.

NCCER applauds your dedication and wishes you the best in your future endeavors.

Sincerely,

Donald E. Whyte
President, NCCER

Enc.

3600 NW 43rd St, Bldg G ○ Gainesville, FL 32606 ○ P 352.334.0911 ○ F 352.334.0932 ○ www.nccer.org

NATIONAL CENTER FOR CONSTRUCTION EDUCATION AND RESEARCH
BUILDING TOMORROW'S WORKFORCE
3600 NW 43rd St, Bldg G • Gainesville, FL 32606
P 352.334.0911 • F 352.334.0932 • www.nccer.org
Affiliated with the University of Florida

August 27, 2009

Official Transcript

Sample Student
National Center for Construction Education and Research
3600 NW 43rd St Bldg G
Gainesville, FL 32606

Current Employer/School:

Card #: 2781481

Course / Description		Instructor	Training Location	Date Compl.
00101	Basic Safety	Don E Whyte		1/1/2001
00102	Basic Math	Don E Whyte		1/1/2001
00103	Introduction to Hand Tools	Don E Whyte		1/1/2001
00104	Introduction to Power Tools	Don E Whyte		1/1/2001
00105	Introduction to Blueprints	Don E Whyte		1/1/2001
00106	Basic Rigging	Don E Whyte		1/1/2001
26101-02	Electrical Safety	Don E Whyte		5/1/2001
26102-02	Hand Bending	Don E Whyte		5/1/2001
26103-02	Fasteners and Anchors	Don E Whyte		5/1/2001
26104-02	Electrical Theory One	Don E Whyte		5/1/2001
26105-02	Electrical Theory Two	Don E Whyte		5/1/2001
26106-02	Electrical Test Equipment	Don E Whyte		5/1/2001
26107-02	Introduction to the National Electrical Code	Don E Whyte		5/1/2001
26108-02	Raceways, Boxes, and Fittings	Don E Whyte		5/1/2001
26109-02	Conductors	Don E Whyte		5/1/2001
26110-02	Introduction to Electrical Blueprints	Don E Whyte		5/1/2001

Donald E. Whyte
President, NCCER

Trade Terms Introduced in This Module

Baseload: Energy used to power things in the home such as lighting and appliances that are used year round. It does not include energy used to heat or cool the home.

Blower door: A variable-speed fan and its controls that is placed in the door of a home. It is used to detect air leaks into or out of a home.

British thermal unit (Btu): The amount of heat energy needed to raise the temperature of one pound of water 1°F. Heat loss and heat gain is expressed in the number of Btus a house gains or loses in an hour (Btu/h). Furnaces and air conditioners are sized in Btus per hour.

Compact fluorescent lamp (CFL): A fluorescent lamp that can be screwed into an incandescent light bulb socket. Compact fluorescent lamps use far less energy than standard light bulbs and they have a much longer life.

Conditioned air: Air within a home that has been heated or cooled, humidified or dehumidified, or cleaned of dirt or other contaminants.

Cubic feet per minute (cfm): A term used to measure airflow.

Energy Star®: A program of the U.S. Environmental Protection Agency. Products that meet Energy Star® requirements are energy-efficient. Use of a rated appliance will result in energy savings for the homeowner as well as reductions in air pollution brought about by more efficient use of energy.

Equilibrium: A condition in which opposing forces cancel each other, resulting in balance.

Exfiltration: The loss of conditioned air from a home to the outdoors.

Heat gain: Heat that enters a home through walls, windows and doors, the roof and through openings to the outside. Home weatherization can reduce heat gain, reducing the home's cooling load.

Heat loss: Heat that is lost from a home through walls, windows and doors, the roof and through openings to the outside. Home weatherization can reduce heat loss, reducing the home's heating load.

Infiltration: The unwanted entry of air into a home through leaks in the exterior of the home.

Infrared camera: A device that shows a thermal image of the home. The color of the thermal image indicates where the home is losing or gaining heat.

Light-emitting diode (LED): A very efficient but costly type of light bulb that uses about one-fifth the energy of an incandescent light bulb, and about half the energy of a compact fluorescent lamp.

Low-E glass: A type of window glass (low emissivity) with a thin, clear metallic coating on the inside of the pane. Low-E glass reduces heat loss and heat gain. In the summer, the coating reflects radiant heat and prevents it from entering the home. In the winter, the same coating prevents heat loss from the home by retaining radiant heat energy.

Radon: A colorless, odorless radioactive gas that is formed by the breakdown of uranium in soil and groundwater. Prolonged exposure to radon can increase the risk of lung disease.

Return air duct: The air duct in a home that returns air within the home to the furnace or air conditioner where it is conditioned.

Rim joist: A board that encloses the ends of the floor joists (supports).

R-value: A number, such as R-11 or R-25, used to indicate the ability of insulation to resist the flow of heat. The higher the R-value, the better the insulating ability.

Stack effect: A condition in a home where warm air rises to upper floors where it leaks into the attic and out of the home. The stack effect functions in a manner similar to a chimney.

Supply air duct: The air duct in a home that carries conditioned air to rooms within the home.

U-value: A measure of a window's ability to resist heat flow. The lower the U-value, the better the window.

Vapor barrier: A barrier placed over insulation that stops water vapor from passing through the insulation and condensing on a cold surface.

Additional Resources

This module is intended to present thorough resources for task training. The following references are suggested for further study. These are optional materials for continued education rather than for task training.

Insulate and Weatherize. Newtown, CT: Taunton Press.

Insulating Materials. Basel, Switzerland: Birkhauser Publishers for Architecture.

Thermal Insulation Building Guide. Malabar, FL: Krieger Publishing Company.

Figure Credits

Flir Systems, Module opener, 101F17, 101F18

U.S. Environmental Protection Agency, 101F04

Drheet.com, 101F05

U.S. Department of Energy, Office of Energy Efficiency and Renewable Energy, 101F10–101F13, 101F22, 101F26, 101SA04, 101F35, 101SA06

Topaz Publications, Inc., 101F14, 101F16, 101F20 (blanket insulation), 101F21, 101F33

Copyright 2006 CertainTeed Corporation, used with permission, 101F19, 101F20 (batt insulation), 101F23

Johns Manville, 101F24

Dow Chemical, 101F25

DAP Products Inc., 101F27

Minnesota Department of Commerce, 101F28, 101F29

Horizon Energy Systems, 101SA03

P3 International, 101SA05

OSRAM SYLVANIA, 101F36

Courtesy of Honeywell International Inc., 101SA07

Index

A

Air leaks
    doors, 1.20
    duct systems, 1.21
    infiltration and exfiltration, 1.4–1.6
    locating, 1.7–1.14
    sealing, 1.17–1.18
American Recovery and Reinvestment Act (ARRA), 1.1
Appliance energy audits, 1.1–1.2
Appliances
    age and condition issues, 1.1–1.2
    baseload energy use, 1.22–1.23
    checking power use, 1.25
Asbestos, 1.9
Attic
    air leaks, 1.9–1.11
    insulation, 1.8, 1.23

B

Balloon frame construction, 1.7
Baseload
    defined, 1.1, 1.21–1.22, 1.37
    elements comprising the, 1.22–1.23
Blower door, 1.1, 1.8, 1.9, 1.12–1.14, 1.37
British thermal unit (Btu), 1.2, 1.37
Btu. *See* British thermal unit (Btu)
Buglione, Steve, 1.32
Building shell tightness, 1.2

C

Carbon monoxide, 1.7
Careers in weatherization
    advancement opportunities, 1.27
    Buglione, Steve, 1.32
    crew chief, 1.25–1.26
    energy auditor, 1.26
    NCCER training for, 1.26–1.27
    technician, 1.25
Caulk, 1.18
CFL. *See* compact fluorescent lamp (CFL)
Cfm. *See* cubic feet per minute (cfm)
Chimneys, 1.4
Compact fluorescent lamp (CFL), 1.23, 1.24, 1.37
Conditioned air, 1.4, 1.37
Contren® Learning Series, 1.26, 1.27
Cooling percent of baseload, 1.22
Craft instructor, 1.27
Crew chief, weatherization, 1.25–1.26
Cubic feet per minute (cfm), 1.12, 1.37

D

Dishwashers, baseload energy use, 1.23
DOE. *See* Energy Department, U.S. (DOE)
Doors, heat loss and gain through, 1.18–1.21
Duct systems
    ideal, 1.4, 1.6
    sealing and insulating, 1.21

E

Electronics baseload energy use, 1.22
Energy audit equipment, 1.1
Energy audit inspection areas
    air infiltration, 1.4–1.6
    building shell tightness, 1.2
    equipment condition, 1.1–1.2
    heat loss and gain, 1.2–1.4
    home health and safety, 1.1
    home lighting, 1.2
Energy auditor
    educating homeowners, 1.24
    responsibilities, 1.1, 1.26
Energy costs, 1.1
Energy Department, U.S. (DOE), 1.1
Energy–recovery ventilator (ERV), 1.8
Energy Star®, 1.22–1.23, 1.25, 1.37
Energy tax credits, 1.1
Energy use
    household breakdown, 1.22
    seasonal vs. baseload, 1.21–1.22
Environmental Protection Agency (EPA), 1.2
EPA. *See* Environmental Protection Agency (EPA)
Equilibrium, 1.2, 1.37
ERV. *See* energy–recovery ventilator (ERV)
Exfiltration, 1.4, 1.37

F

Flexible insulation, 1.15
Forced–air duct systems, 1.5

G

Glass
    heat loss and gain, 1.18–1.21
    low–E, 1.19, 1.20, 1.37
    U–values, 1.20, 1.21, 1.37

H

Health, leaking ducts effect on, 1.7
Heating baseload energy use, 1.22
Heat loss and gain
    air leaks, 1.4–1.14
    defined, 1.37
    doors and windows, 1.18–1.21
    insulation, inadequate, 1.12–1.14
    overview, 1.2–1.4
    radiant barriers for, 1.23
Heat pump water heater, 1.26
Heat–recovery ventilator (HRV), 1.8
Home construction
    balloon frame, 1.7
    heat loss and gain factors, 1.2
Home health and safety audit, 1.1–1.2
HRV. *See* heat–recovery ventilator (HRV)

I

Infiltration, 1.4–1.6, 1.37
Infrared camera, 1.1, 1.12–1.14, 1.37
Insulation
   adding, choices for, 1.14–1.18
   air leaks and, 1.12–1.18
   R–value, 1.14–1.15
   seasonal effects, 1.14
   types of
      flexible, 1.15
      loose–fill, 1.16
      rigid foam board, 1.15, 1.16
      spray foam, 1.16, 1.17
      spray–in–place, 1.16, 1.17

L

Lawrence Berkeley National Laboratory, 1.22
Lead–safe training, 1.1, 1.2
LED. *See* light–emitting diode (LED)
Light–emitting diode (LED), 1.23, 1.37
Lighting, household, 1.2, 1.23, 1.24
Loose–fill insulation, 1.16
Low–E glass, 1.19, 1.20, 1.37

N

National Center for Construction Education and Research (NCCER)
   accreditation process, 1.27
   overview, 1.26
   standardized training, 1.26–1.27
   training credentials, 1.33–1.36
   website address, 1.27
National Center for Construction Education and Research (NCCER) Accredited Assessment Centers, 1.27
National Center for Construction Education and Research (NCCER) Accredited Training Sponsors, 1.27

P

Power meters, 1.25
Programmable room thermostat, 1.27
   installing, 1.24

R

Radiant barriers, 1.23
Radon, 1.7, 1.37
Refrigerant, 1.24
Refrigerators baseload energy use, 1.22–1.23
Return air duct, 1.4–1.6, 1.37
Rigid foam board insulation, 1.15, 1.16
Rim joist, 1.17, 1.37
Roofs, energy–efficient, 1.21
Rope caulk, 1.18
R–value, 1.14–1.15, 1.16, 1.37

S

Seasonal energy use, 1.21–1.22
Spray foam insulation, 1.16, 1.17
Spray–in–place insulation, 1.16, 1.17
Stack effect, 1.4, 1.6, 1.8, 1.37
Supply air duct, 1.4–1.6, 1.37

T

Tax credit qualifications, 1.1
Technician, weatherization, 1.2, 1.25
Training
   lead safety, 1.2
   NCCER
      credentials, 1.33–1.36
      standardized, 1.26–1.27

U

U–value, 1.20, 1.21, 1.37

V

Vapor barrier, 1.15, 1.37
Ventilation, 1.7
Ventilators, recovery, 1.8
Vermiculite, 1.8
Visual inspections, 1.8, 1.12

W

WAP. *See* Weatherization Assistance Program (WAP) (DOE)
Washing machines baseload energy use, 1.23
Water heaters baseload energy use, 1.22, 1.23, 1.24
Weatherization
   introduction, 1.1
   steps in
      adding insulation, 1.14–1.18
      air leaks, sealing, 1.17–1.18
      ducts, sealing and insulating, 1.21
      roof upgrades, 1.21
      window and door upgrades, 1.18–1.21
   summary, 1.28
Weatherization Assistance Program (WAP) (DOE), 1.1
Weatherization programs, 1.1, 1.20
Weatherstripping, 1.18
Windows, heat loss and gain through, 1.18–1.21

CONTREN® LEARNING SERIES — USER UPDATE

NCCER makes every effort to keep these textbooks up-to-date and free of technical errors. We appreciate your help in this process. If you have an idea for improving this textbook, or if you find an error, a typographical mistake, or an inaccuracy in NCCER's Contren® textbooks, please write us, using this form or a photocopy. Be sure to include the exact module number, page number, a detailed description, and the correction, if applicable. Your input will be brought to the attention of the Technical Review Committee. Thank you for your assistance.

Instructors – If you found that additional materials were necessary in order to teach this module effectively, please let us know so that we may include them in the Equipment/Materials list in the Annotated Instructor's Guide.

Write: Product Development and Revision
National Center for Construction Education and Research
3600 NW 43rd St, Bldg G, Gainesville, FL 32606

Fax: 352-334-0932

E-mail: curriculum@nccer.org

Craft _____ Module Name _____

Copyright Date _____ Module Number _____ Page Number(s) _____

Description
_____
_____
_____
_____
_____

(Optional) Correction
_____
_____
_____
_____

(Optional) Your Name and Address
_____
_____
_____